Fabiana Gama Chimes

The Influence of the Pokémon Cartoon

Fabiana Gama Chimes

The Influence of the Pokémon Cartoon

The Influence of the Pokémon Cartoon on the
Development of Learning in Botany in
Elementary School II

ScienciaScripts

Imprint

Any brand names and product names mentioned in this book are subject to trademark, brand or patent protection and are trademarks or registered trademarks of their respective holders. The use of brand names, product names, common names, trade names, product descriptions etc. even without a particular marking in this work is in no way to be construed to mean that such names may be regarded as unrestricted in respect of trademark and brand protection legislation and could thus be used by anyone.

Cover image: www.ingimage.com

This book is a translation from the original published under ISBN 978-613-9-67618-7.

Publisher:
Sciencia Scripts
is a trademark of
Dodo Books Indian Ocean Ltd. and OmniScriptum S.R.L publishing group

120 High Road, East Finchley, London, N2 9ED, United Kingdom
Str. Armeneasca 28/1, office 1, Chisinau MD-2012, Republic of Moldova, Europe
Printed at: see last page
ISBN: 978-620-8-17510-8

SUMMARY

DEDICATORY

I thank God in the first place, because without him I wouldn't have had the strength for this long journey; I thank my parents and siblings, my teachers for the knowledge they passed on, my friends who helped me and made this journey lighter.

THANKS

No battle is won alone. Throughout this struggle, some people have been by my side and have walked this path like true soldiers, encouraging me to seek my victory and achieve my dream.

First of all, I thank God for everything, for lifting me up every time I needed it and practically sustaining me throughout this journey full of obstacles, losses, achievements and for the opportunity to be conquering something I wanted and planned so much.

I would like to thank my parents Ronaldo Chimes and Marli Gama Chimes for all their support, patience, education, love, support and understanding when I didn't have time to spend hours studying, for teaching me to have dignity and to be a better person every day and for making me so happy, for teaching me the greatest values, to fight and pursue my goals always with my feet on the ground, always showing me the right way to go.

Mom, Dad, if I could come back to life at another time and had the opportunity to choose my parents, I would choose you, because I'm sure you're the best in the world. I'm very proud to have you as my parents.

I would like to thank my siblings Silvia R. G. Chimes and Edson G. Chimes for all their support during this journey, for their friendship, for the laughs, the tugs on the ear, the affection, the love, for all their help at times when I thought of giving up. I love you! I would like to thank your spouses for becoming my brother and sister. My beloved niece Eshiley C. C. Domingos for all her affection, jokes, love, for being my "xodó", for being so special in our lives and for supporting me in my dream.

I would like to thank my great friend Amanda P. Lopes for always being there for me over the years, for her companionship, for her sincere friendship, for her ear tugs, for her help when I needed it most, for being there when I wanted to cry, for laughing for hours, for always telling the truth, for her strength and understanding, for encouraging my dreams and pushing me not to give up, for your respect, for the trust you placed in me, for your affection, for your companionship during the work and projects that took us late into the night, for the late-night chats, for the ideas for new projects, for the trips to rest from the stress of the work. Life puts special people in our path and surprises us with the capacity for a friendship that is so loyal, important, rewarding (yes, I'm very grateful for your friendship), honest, and a person of great value, for always supporting me and believing in my potential, you are a fundamental part of this trajectory. Thank you for everything and for making this long journey lighter!

Thank you so much to the friends I've made over the years: Thaisa Olivera and Bernardo you live in my heart, thank you for all your support; Priscilla Dàlia for the laughs, sharing stresses, for your friendship; Graziele Silva; Caroline Ramos (for her sincere friendship); Moisés Fiùza (who has put up with me since childhood); Dirlane Fontes; Thamiris Dias; Joice Neves; Vanessa Marques; Grabrielle Freitas thank you so much for your strength, friendship, affection, support, laughs, and so many others I haven't mentioned but who are part of this journey. Thank you for making the walk so much fun.

I would like to thank my great friend Filipe Araùjo for all his support, affection, trust, strength, tugs on the ear to get me to sleep, for his encouragement, for agreeing to be present at the most important moments of this journey. Thank you for your friendship!

I would like to thank my friend Ademar F. da Silva for his help throughout the course, for his teachings and patience, for his strength, for being so helpful, a friend, for loving what he does and passing it on every day.

I would like to thank the librarian, Tânia Maria de Carvalho Pereira, for all the attention, sympathy and patience she always showed, always being helpful and helping me with countless bibliographies that I didn't even remember.

I would like to thank my friend Arion T. Aranda for his friendship, encouragement, patience, professionalism and shared knowledge. When I grow up, I want to be just like you, and the entire Pedra Branca Project team (Thaise, Anderson, Vinicius and Virginia).

I would like to thank all the teachers I have had the chance to learn from and get to know over the years. In particular, I would like to thank Professor Adriana Cabrai for her friendship, for the trust she has placed in me, for helping me to love biology more and more, for her joy and dedication in the classroom, for being the professional I want to be like one day, not just for her training, but for the example she has set for my life; Professor Izabel Côrtes for her friendship, affection, understanding, patience, for the love she has for her profession and for transmitting this very well to us, thank you for being my co-supervisor; Professor Alicelena Bustamante for accepting me as her monitor for so many years and believing in my work; Professor Luis Henrique Pereira for his friendship, affection, for encouraging me in my career and in projects; the coordinator Luciana Dolinsky who always helped me with my doubts, for the friendship we built, for always being so helpful and always present, doing her coordinating role in an exemplary way.

I would like to thank my advisor André Costa Ferreira for his availability, attention and understanding, and for all his help in making this monograph a reality.

Thank you to everyone who has contributed so far, I promise you that this is just the beginning.

"Let your efforts defy impossibilities, remember that the great things of man have been achieved from what seemed impossible."

Charles Chaplin

SUMMARY

For many years, science teaching has been mainly dedicated to the transmission of currently accepted knowledge, with little emphasis on the actual procedures for constructing science (Silva *et al*, 2009). Based on the construction of knowledge through play, quality education can be achieved by meeting the interests and needs of children in elementary school, together with the media. This work aims to analyze the different forms of influence of cartoons, identifying those Pokémon types with "plant" characteristics. The knowledge of 7th grade elementary school students was assessed through a questionnaire, in which it was observed that 80% of the students, through the influence of the cartoon, are able to distinguish and identify the relationship between the character and the real thing. The drawing in question can help the students' development and interest in learning botany.

Keywords: Influence of Animation; Playfulness; Constructed Concepts; Elementary II; Pokémon; Plant.

1 INTRODUCTION

Science teaching is mainly dedicated to transmitting currently accepted knowledge, with little emphasis on the actual procedures for constructing science (Silva *et al*, 2009).

The approach to learning science is transformed into a collection of difficult names to be memorized en masse. The approach to biological content without proper contextualization makes it difficult to learn the core concepts of science teaching. Thinking about how this content is taught, it can be described that, for the most part, the methodology that predominates in the classroom is expository, characterized by the systematized and linear presentation of content to students (Silva *et al*, 2009).

Elementary school students say that science lessons should contain many more experiments, visits to museums and other environments, and other resources that are more attractive, such as films, drawings, comics etc., without the number of strange terms seen in the materials used in class. With regard to the teaching of botany in primary school, the subject is approached by means of a list of scientific names and words that are totally isolated from reality, used to define concepts that are difficult for students to understand. Botany teaching strategies are still linked to expository teaching, which increases the stigma of the subject (Silva *et al*, 2009).

Botany concepts are taught in a discouraging and unpleasant way, without observation or direct interaction with plants. The students behave like mere listeners and the knowledge passed on by the teachers is not even assimilated or learned, but simply memorized for a short period of time, which does not characterize robust learning.

1.1 - National Curriculum Parameters (PCN)

The National Curriculum Parameters (PCN) aim to help in the execution of classroom work, sharing efforts on a daily basis to ensure that children master the knowledge they need in elementary school and become fully recognized and aware of their role in society.

The PCNs are a quality benchmark for elementary school education throughout the country. Its function is to guide and guarantee the coherence of investments in the educational system, socializing discussions, research and recommendations, subsidizing the participation of Brazilian technicians and teachers, especially those who are more isolated, with less contact with current pedagogical production (PCN, 1998).

In the 1990s, Brazil took part in the World Conference on Education for All in Jomtien, Thailand, convened by UNESCO, UNICEF, UNDP and the World Bank. At this conference, the New Delhi Declaration was signed, which resulted in consensus positions in the fight to satisfy basic learning needs for all, making primary education universal (PCN, 1998).

The contribution that science teaching provides is a process of transmitting scientific truths, without the possibility of discussing the contradictions and ideological positions that exist in the process of scientific production.

Throughout the history of education, it has been observed that science teaching has only been compulsory for the junior high school grades since 1971. Law 5.692 made the subject compulsory for the eight grades of the former elementary school: the current nine years of elementary school. Despite four decades of compulsory education, many schools and teachers are still not working on the subject in the 21st century. The reasons are varied, ranging from a lack of resources and knowledge to the belief that science teaching is not important.

The quality of the subject was defined by the amount of content covered. The main resource for study and assessment was the questionnaire, which students had to answer by focusing on the ideas presented in class or in the textbook chosen by the teacher. Proposals to renew science teaching have become necessary to meet the needs of the curriculum due to the advance of scientific knowledge and pedagogical demands. Science has been useful in this teaching proposal, as knowledge of theories from the past can help to understand the conceptions of present-day students, as well as being relevant learning content. According to the PCNs, Biology is included in the area of Natural Sciences, Mathematics and their Technologies, and its main objectives are to work on basic concepts, analyze the processes of scientific research and their social implications (PCN, 1998).

Botany, from the Greek "botàne", meaning plant, is the part of biology that studies, groups and classifies plants into categories according to their similar characteristics. It could be considered a science that is better understood and accepted in the classroom, but this is not what we see. This situation is probably due to the way in which botany content is transmitted, without any link to the student's daily life and reality (PCN, 1998).

(...) learning about the diversity of life can be made meaningful to students through opportunities to come into contact with a variety of species that they can observe, directly or indirectly, in real environments, considering them as one of the components of wider systems. (...) which should provide students with knowledge about the forms and functions of the body related to the habits and habitats of living beings, helping to form a broad and interesting panel about life on Earth. The sources of information to be worked on with the students will be real images of environments and descriptive and narrative texts about environments and living beings, including excerpts from historical texts by naturalists of the past. (...)) living things can be examined in the environment of a garden, square or park; a cultivated or abandoned field, as mentioned above; houses, apartments, streets and rivers in cities; certain aquatic and terrestrial environments; The characterization of the herbaceous, shrubby and arboreal strata present in different environments represents a significant advance in the recognition of the plant components of landscapes, allowing for an interesting description of vegetation and the identification (....) of different stages in the process of recomposing the natural environment. The description and comparison of significant plants from certain studied environments is also important, and provides a repertoire for recognizing the existence of plants that do not have seeds and others that do (BRASIL, 1998, p. 69-70).

In addition to what the Curriculum Guidelines for Primary Education say about science teaching, botany comes up, with the content of plant diversity, then morphology and physiology, as well as anatomy and evolution.

1.2 - Using play for learning

A playful activity is an entertainment that gives pleasure and amuses people, and is an integral part of every human being's life. Games and toys are part of children's childhood, where reality and make-believe intertwine. Play should not just be seen as fun, but as an activity of great importance in the teaching-learning process. We can say that play functions as an integrating link between the motor, cognitive, affective and social aspects, so from play, the facility for learning, social, cultural and personal development develops and contributes to a healthy physical and mental life. Playing is a natural impulse of the child, which, when combined with learning, makes it easier to learn due to the spontaneity of playing in an intense and total way. The problem that will be analyzed seeks to understand the importance of animations as effective aids in the construction of children's knowledge through the necessary stimuli in the production of their learning (Malaquias & Ribeiro, 2013).

Based on the construction of knowledge through play, quality education can be achieved that meets the interests and needs of elementary school children. The knowledge acquired from lived experiences becomes significant, exploring and involving students in their own imaginary world, as animations help them to invent and reinvent in order to better approach botany (Corrêa, 2012).

Teachers are gradually seeking information and enriching their experiences in order to understand play and how to use it to help build children's learning. Anyone who works in children's education should know that we can always develop a child's motor skills, attention and imagination by playing with them. Play is the teacher's partner (MALUF, 2003, p. 29).

The traditional classroom approach to botany, with a blackboard, pen, book, exercises and so on, is not sufficient for a full understanding of the subject. Given that a large number of schools now have video rooms or laboratories equipped with computers, it is easy to implement some activities for students where the content can be developed using animations to improve understanding of the content taught in class (Castilho & Ricci, 2004).

According to Piaget, changes in cognitive structure are part of an individual's development. And this development arises from action, which is triggered by stimuli originating in the external environment or by an inner need to satisfy one's own desires. The search to satisfy these needs, which can even be an answer to an inner question, generates action. Once this need has been satisfied, others arise, and the action that generates a balance is unbalanced by the transformations that appear in the world, whether external or internal, and each new behavior will work not only to re-establish the balance, but also to tend towards a more stable balance than that of the stage prior to this disturbance (Piaget, 2002).

Imagination and curiosity also make a difference when it comes to learning science. Scientific activities become important and stimulating when they are able to excite our curiosity. Through our imagination, our thoughts begin to grasp the unknown, seeking an explanation for our enigmas. Curiosity then serves as a guiding principle for activities that would not have the same meaning if they were merely bureaucratic and carried out with the aim of fulfilling obligations. It can be said that curiosity arises from the unknown that can somehow be grasped by the imagination.

"There is an educational dimension to cartoons, especially if we consider the active aspect of the values that can be built up when the child interacts with them. This, on the other hand, cannot be confused with a type of directive pedagogy, where the cartoon brings values and models that are determined to be copied by the child, in order to affect and shape their conduct" (SALGADO, 2005, p. 8).

1.3 - The history of television - digital media

The creation of cinema in 1895 and television (TV) in 1926, a brilliant and spectacular innovation, marked the beginning of image and sound transmissions. From that time to the present day, this technology has enabled a communication strategy. With its origins in Chinese shadow theater, animated cinema underwent important transformations until it reached TV. With the integration of digital technology, cinema has undergone a paradigm shift, and in the same way, the animation genre has been reformulated. The trajectory of animation reveals a story that involves important technical advances, with the focus being on children, and then the growing adherence of a heterogeneous audience, extending from children to adults. Cartoons have many ways of dialoguing with the universe of children, and after appropriating the content transmitted by them, young people develop their expressions and emotions, which can be fundamental for the elaboration of their conflicts and the resolution of problems in a creative way (Fossatti, 2009).

Radio also came to disseminate news and interact with our minds with its radio soap operas. It all began in 1863 when, in Cambridge (England), James Clerck Maxwell theoretically demonstrated the probable existence of electromagnetic waves. The beginning of radio propagation came in 1887, through Henrich Rudolph Hertz (1857-1894). The first official radio broadcast in Brazil was a speech by President Epitàcio Pessoa, in Rio de Janeiro, during the celebration of the centenary of Brazil's Independence, on September 7, 1922. The speech took place at an exhibition in Praia Vermelha, Rio de

In 1931, PRB 9 - Ràdio Record de Sao Paulo was founded, followed by other radio stations that became popular in Brazil, such as Tupi, Ràdio Globo, Ràdio Nacional and many others (microfone.jor.br, 2008).

Most people think that it was the Scotsman Alexander Graham Bell who created the telephone, which is not true. The American Congress, in resolution 269 of 2002, recognized that the official inventor of the telephone was the Italian Antonio Meucci, who in 1860 had already described the device in a New York Italian-language newspaper under the name "teletrophone" (History of everything). Until recently, the best known and most common way of connecting the user to the telephone exchange worldwide was the metal wire, but there are other ways. The radio system has long been used to connect subscribers who are far from telephone exchanges or in places that are difficult to reach. Just as we receive TV and radio signals transmitted on frequencies, we also receive voice signals on frequencies, replacing the metal wire. With technological developments and the growing demand for ease of communication, Cellular Mobile Telephony emerged. In 1984, the analysis of cellular technology systems began in Brazil, and the American standard, analog AMPS, was defined. The first city to use the service was Rio de Janeiro, in 1990 (projetos.unijui.edu.br, 2008).

Today there are more cell phones than inhabitants in the country. With the creation of the internet at the end of the 1960s, in the midst of the Cold War, thanks to an initiative by the American Department of Defense, the combination of cell phone technology and the internet has allowed us to have an important and useful tool, a pocket computer that we can use to play video games, read digital books, consult Google, make bank transfers and many other things that only this 21st century technology allows us to do.

Cartoons also encourage interaction between children and the people around them - parents, siblings, friends, teachers, etc. The identification is not limited to the moment when the animation is watched, but extends to conversations in the car, at lunch, the characters' outfits, the decorations at the birthday party, the games at school. In other play processes, drawings represent an interesting and fun opportunity to enter the vast territory of children's imaginations (Fossatti, 2009).

In Brazil we've had great attractions for children, such as Vila Sèsamo (1970s) being shown in black and white in a co-production by TV Cultura and Rede Globo. In the USA, it helped children to become literate and to live together. Another hit in the country was the Japanese series Nacional Kid (1960s), where he came from the planet Andromeda in the Alpha Centauri star system to learn about Earth culture. When he arrived, he started defending planet Earth from creatures who wanted to take it over. It was a huge national hit. In the 80s and 90s, many cartoons were popular with children and influenced generations:

• Captain Planet (1990s), the hero was born from the combination of five earth elements: earth, fire, water, wind and heart - and also became the rings of his helpers who, when raised and said aloud, asked for his help, which was answered with the phrase "By the union of god powers, I am Captain Planet".

• Thundercats (1980s) the series used as a basis for a wide range of stories that were a mix of science fiction and fantasy in a good traditional tale of good versus evil, featuring more and more allies and villains in the world of the Thundercats. Futuristic technology was as central to the series as magic and myth, but even in the midst of all this action, it emphasized the importance of moral values in solving problems.

• He-Man, the cartoon was born in 1983 and the story was about the struggle of He-Man and his friends to defend the planet Eternia and the Castle of Grayskull from the evil forces commanded by the villain Skeleton.

• She-Ra, the princess, was born in 1985, as a bet on the female audience due to the success of He-Man. In the story, it's her twin sister who was captured at birth.

• Cave of the Dragon (1980s), the cartoon told the story of six youngsters in an amusement park embarking on a rollercoaster called Dungeons & Dragons. During the ride, a portal opened and transported the children to another world, where they appeared wearing different clothes and carrying magical weapons from someone who introduced himself as the Master of Wizards. During the episodes, the youngsters went on various adventures in their quest to return home.

• Bob's Fantastic World (1990s) was a cartoon about the daily thoughts of Bobby Generic, a four-year-old boy with an extremely fertile imagination. The episodes were based on his childish perceptions, and also featured his family: his father, Howard; his mother, Martha; and his two brothers, Derek and Kelly. His friend Jackie, his Uncle Ted, Aunt Ruth and his dog also make constant appearances.

• Knights of the Zodiac (1990s), the story featured five mystical warriors called Knights, who fought wearing sacred armor based on the constellations that protect each one. Their mission was to defend the reincarnation of the Greek goddess Athena in her battle against other Olympian gods who wanted to take over the Earth.

- Care Bears (1990s), the cartoon told the story of a family of caring bears who help people cultivate good feelings, and protect the Earth from the shadows of evil and the dreaded villain Ice Cream Heart, who tries at all costs to put an end to love.

- The first version of the cartoon appeared in 1968. In the episodes, the competitors sought the world title of World's Craziest Runner. The characters included: Dick the Trickster, Muttley, Penelope Charming, Rufus the Lumberjack, Clyde, Peter Perfect and Sergeant Bombarda.

- The Jetsons (1990s), the theme of the cartoon was the Space Age and each episode introduced to most people's imaginations what everyday life would be like in the future: flying cars, suspended cities, automated work, all sorts of household appliances and entertainment devices, robots as household helpers, and much more.

- Dragon Ball (1980s), the story of the cartoon tells of Goku's life, from his adventures as a child to becoming a grandfather. Throughout his life, he takes part in many battles with stronger and stronger opponents, always seeking to increase his strength. On his travels, he meets various people, often forming strong friendships with them, who help him defeat the villains of the story.

- Our Gang (1980s), the cartoon told the adventures of a gang of twelve anthropomorphic pre-adolescent animals, led by the moose Montgomery, always valuing team spirit and companionship.

- Tom & Jerry (1940s), the cartoon was based on Tom's failed attempts to capture Jerry, and the chaos and destruction that ensues (Weekly Guide, 2015).

And many others that marked the childhood of many who are now adults and still follow with nostalgia. But not everything turns out the right way, and the very cartoon to be discussed in this article also ended up having some problems in its early days.

On December 16, 1997, an episode of Pokémon aired in Japan. Approximately thirty minutes later, almost 700 children ended up in hospital. The episode called "Electric Soldier Porygon" became a legend because its premise was quite innocent: the heroes, upon discovering a problem in a Poké Ball transfer machine at a Pokémon Center, decide, with the help of a scientist and his Porygon, to enter the machine and find out what's wrong in the digital world. What caused all the problems was the animation technique used in this episode. At one point, about twenty minutes into the episode, Pikachu uses his thunder shock to blow up some missiles. As these are virtual missiles, and they are in cyberspace, a normal explosion would not go down well. The animators used a fast and bright technique that flashed red and blue lights on the screen, to make the explosion look "virtual". Immediately, children all over Japan began to have various problems. Some fainted, or had blurred vision. Others felt dizzy or nauseous. In extreme cases, epileptic seizures and even temporary blindness were reported. It is impossible to know the exact number of children affected by the episode, as most of the cases were minor, but a total of 685 children (375 girls and 310 boys) were put in ambulances and taken to hospital. Most recovered quickly, some within minutes, but a small number were diagnosed with epilepsy, which was caused by the bright image flashing too quickly. It was a disaster for children's animation in Japan, shares in the Pokémon brand and Nintendo plummeted, and the cartoon was off the air for four months while producers and health professionals tried to find out what had caused the problem in such a large number of children (Neogamer, 2011).

Soon after this episode, the cartoon had undergone several changes, and since then Japanese television and cinema have created cartoons and films that are hits all over the world and influence generations.

1.4   - Anime History

Anime is a term that defines cartoons of Japanese origin and also the elements related to these cartoons. In Japan, anime refers to animation in general. Anime is traditionally drawn by hand. However, with the development of technological animation resources, especially since the 1990s, many have been produced on computers. The first anime in history was created in 1907 in Japan. It was a story whose main character was a boy sailor (Luyten, 2004).

The themes covered in anime are very varied (drama, fiction, horror, adventure, psychology, romance, behavior, mythology, etc.). Another important feature of today's anime is the occurrence of technological elements in the storylines. Anime is very successful in Japan and in many countries around the world, including Brazil. The animations are made for the cinema, television and comic books. An example of anime that has been and still is successful in Japan and around the world is Pokémon, as well as many others (Luyten, 2004).

1.5   - History of Pokémon

Pokémon is the brainchild of Japanese programmer Satoshi Tajiri and his friend, artist and designer Ken Sugimori. After its first games, several others were produced, a total of 43, and the series expanded into several manga, an official card game, an anime, now in its 19th season[a] , 6 generations and 12 films already released, plus a 13th in production. Pokémon became a landmark in pop culture from the 1990s until 2003 and sales of its games exceeded 180 million units worldwide, which led the series to become Nintendo's second best-selling game and also the world's best-selling game, both times second only to the Mario Bros. series. The origin of the entire series is the video games made for Nintendo consoles (Pokémon Best World, 2010).

The main features of the Pokémon games are the need to collect different monsters and the option to choose which ones will be part of the player's group and how they will be trained. The games have often been considered innovative in terms of connectivity, since Pokémon Red, Blue & Green made it possible to link video games by connecting two Game Boys via the Game Link cable and allowing players to exchange Pokémon and battles. Created by the company Game Freak, the Pokémon games were intended to interact with players and make them interact with others, battling and exchanging Pokémon from one version to another. The Pokémon video games known as the original RPGs always follow a set script, which is to capture as many Pokémon as possible, become the champion by beating the greatest trainers in the region and also defeat a criminal organization that wants to take over the world. After the first games, the creators made an anime, which marked the beginning of Pokémon's "invasion" of the West in the late 1990s, and also led to the arrival of other anime such as Dragon Ball and Dragon Ball Z, Sakura Card Captors and Digimon (considered the biggest rival to the Pokémon cartoon). With more than 620 episodes aired in Japan, the Pokémon anime is the fifth longest-running cartoon in the United States, surpassed only by The Simpsons, The King of the Hill, Arthur and South Park. In Brazil, a CD entitled Para *Ser Um Mestre* was released by the Abril Music label, with the songs in Portuguese and all the artists are Brazilian (massarani.com.br, 2010).

When he was young, Satoshi Tajiri's hobby was entomology. When he grew up, Tajiri decided not to go to college and would always quit the jobs his father got him so he could play arcades, until he decided to take a technical course and then created his magazine, known as Game Freak. While working at the magazine, Tajiri met Ken Sugimori, with whom he became friends and worked for a long time. With the growing success of the Nintendo Entertainment System (NES), the two decided to create something innovative for the console. Tajiri turned the magazine into a company, Game Freak, and started working on a game. Released in 1989, the puzzle game Mendel Palace (known in Japan as Quinty) was reasonably successful and marked the beginning of the company's history. The following year, the two decided to create a game for the Game Boy, which had enjoyed great success with Tetris. Seeing the Game Link cable, Tajiri thought of the idea of passing information from one Game Boy to another. Influenced by series such as Final Fantasy and Dragon Quest and associating the idea with metamorphosis, Tajiri created an RPG where monsters could evolve and be passed from one laptop to another. Taking the project to Nintendo, Tajiri, who had the basic idea, and Sugimori, who had the drawings of the monsters, received advice from Shigeru Miyamoto, creator of Super Mario Bros. and The Legend of Zelda, to improve the game, at the time known as Capsule Monster. Production lasted two months. In the meantime, Nintendo was already in decline and Sony was running out of ideas.

the portable. Few at Game Freak believed that the game would be successful and even with a lack of resources, the company's shares were in jeopardy with Pocket Monsters (kassdouglas.blogspot.com.br, 2010).

In February 1996, Pocket Monsters Red & Green was released. Initially, the game was not a success, but as the months went by, more units were sold, reaching the one million copies mark in one year. Nintendo then decided to bring the series to the West. However, it was renamed Pokémon because there was a Mattel series known as Monster in My Pocket and also because it was released shortly after the attack in Japan due to the episode involving Porygon, one of the Pokémon. The series became a smash hit in the United States with Pokémon Red & Blue. The series had a slogan, which is no longer used. In Japan it was known as "Let's Catch Pokémon!", which became the famous phrase "Gotta Catch'em All!" in the United States and became known in Brazil as "Pokémon, Temos que Pegar!". By looking at the Pokémon plants, we can assess the influence that each of them can have on a child and their learning. Cartoons influence those who watch them, the animated world is more attractive in terms of learning (Pokémon Best World, 2010).

Figure 1 - Illustration of some Pokémon characters.

Available at: https://kassdouglas.blogspot.com.br/

1.6   - Pokémon GO - Game

A big hit at the moment is Pokémon GO, a mobile game created by John Hanke and developed by Niantic in partnership with Nintendo and The Pokémon Company. In a short space of time, the game has become a fever worldwide (pokemongobr.club., 2016). Some biological divisions of animals and plants can be observed in the game. By the way, the Pokémon GO game, which uses compatible cell phones, can be used in a playful way to provide greater interaction for students in the classroom. Today, a large number of people have joined in meteorically, making it one of the most popular virtual reality games today.

Figure 2 - Pokémon Go.

Available at: https:// gamedetonado.com.br/

1.7   Justification for the chosen design

1.   The study was based on the cartoon Pokémon, in order to assess the influence that the animation has on the concepts created by students in the 7th grade of elementary school

II.

2.   The theme of the drawing was used as a teaching tool, where we find the biological part throughout the drawing, but we will only focus on the plants. Students have a great deal of difficulty and lack interest in plants because they have so many scientific names, which makes it difficult for them. Botany is not much encouraged in schools, few teachers have the interest and time to approach it more thoroughly and in a way that encourages children to understand plants in a more playful way.

The plant Pokémon have the advantage of generating curiosity and showing the subject in a totally different way, as well as being used as a teaching tool, helping with lessons and attracting students' attention, so that the subject taught becomes more interesting, and can also help develop scientific and critical thinking. Combining drawing with the content of the botany subject may be the best way to attract children's attention and arouse their curiosity.

Identify the similarities between the "little monsters" in fiction and in real life, with the botany content that was covered in class and what was seen in the drawing at home by the students.

Identify the errors and successes in the answers given by the students through the questionnaire.

2. OBJECTIVES

2.1 General

To analyze the different forms of influence of cartoons on the construction of concepts by 7th grade elementary school students in the field of botany, identifying those Pokémon types with "plant" characteristics.

2.2 Specifics

- Specify the importance of using playful techniques in science teaching;

- To administer a questionnaire to 7th graders about the design chosen for study;

- To analyze the concepts that the students have constructed on the theme of botany in the Pokémon cartoon;

- Compare the concepts obtained with the didactic concepts;

- Describe the advantages and disadvantages of using cartoons as a teaching tool.

3. METHODOLOGY

3.1 Analysis of animation concepts

This study began in August 2015 and ended in December 2016. A survey of similar characteristics between the "little monsters" in the Pokémon cartoon and real plants was carried out on the website www.universoanimanga.blogspot.com.br, and identified. This resulted in 19 seasons and 6 generations, with the 19ª season lasting until the present day.

The cartoon features characters who look like plants you've seen somewhere before. And it makes you stop and wonder about the resemblance. By searching for images on the internet, it was possible to find plants that resemble Pokémon. The analysis was carried out by comparing the images and the description of each "little monster", thus making it possible to arrive at the results obtained.

3.3 Description of data collection and sample group

A standardized questionnaire was developed containing closed questions on socio-economic issues and on the Pokémon cartoon. This questionnaire (Appendix 1) contains 15 objective questions, and aims to indicate whether the students are aware of the aforementioned cartoon and to ascertain whether the teaching of botany is corresponding to the initiatives for the possible use of such media as teaching tools.

The questionnaire was administered to 35 students in class 701, 7th year of elementary school II at Colégio Souza Marques, located at Av. Ernani Cardoso, 335, Cascadura, Rio de Janeiro, Brazil. The students were assessed and encouraged to answer the questions in order to verify this influence.

The questions were unbiased as to the gender of the participants, as there is no intention of making an assessment of this type in this work. A private school was chosen. The questionnaire was administered in the classroom, with the help of the teacher and with the permission of the institution involved (Appendix 2).

4. RESULTS AND DISCUSSION

4.1 Pokémon plant survey

By comparing the images obtained on the internet, of which there were 737 analyzed, on the website www.universoanimanga.blogspot.com.br, all the generations were cited in this work, where most of the Pokémon were identified with characteristics of some plant that exists in real life, but many were left out because they had more animal than plant characteristics. In the six generations of Pokémon, there are 95 plant types, 41 of which were discarded because they had more animal characteristics than plant ones.

4.1.1 Seasons and generations of the cartoon

The table below describes each of the 6 generations, the number per generation and the total number of Pokémon plants.

The 6 Generations are made up of:		
Generation	No. of Pokémon	No. of Pokémon plants
1≡ generation	151	12
2≡ generation	99	13
3≡ generation	134	18
4≡ generation	106	19
5≡ generation	155	25
6≡ generation	87	8

Table 1: Description of the generations of Pokémon.

The table below describes each season, episode and number of episodes, of which the 1st[a] season began airing in 1997 and the 19th[a] season to date.

The 19 rounds are made up of:

- 1[a] season - Indigo League (comprising 81 episodes);

- Season 2 - Adventures in the Orange Islands (comprising 35 episodes);

- Season 3 - The Johto Journey (41 episodes);

- Season 4 - Johto League Champions (52 episodes);

- Season 5 - Master Quest (52 episodes);

- Season 6 - Advanced (40 episodes);

- Season 7 - Advanced Challenge (comprising 52 episodes);

- Season 8 - Advanced Battle (52 episodes);

- Season 9 - Battle of the Border (47 episodes);

- Season 10 - Diamond and Pearl (51 episodes);

- 11ª season - Dimensional Battle (comprising 52 episodes);

- Season 12 - Galactic Battles (52 episodes);

- Season 13 - DP - Sinnoh League Winners (comprising 34 episodes); ___________

- 14th season - Black and White (48 episodes);

- Season 15 - Black and White - Rival Destinies (47 episodes);

- Season 16 - Black and White - Adventures in Unova/Adventures in Unova and Beyond (45 episodes);

- Season 17 - The XY Series (48 episodes);

- 18th season - The XY Series - Mission Kalos (45 episodes);

- 19th season - The XYZ Series (number of episodes OPEN, as it is still unreleased in the country).

Table 2: Description of the seasons and episodes of Pokémon.

All the seasons have been broadcast on free-to-air and pay-TV channels such as RedeTV! Rede Globo, Rede Record, Cartoon Network and Tooncast (poke-blast-news, 2016).

4.1.2 Generations and their divisions

- First generation

From Pokémon 01 to 151; The plant characters are:

001 – Bulbasaur   **Bulbo**

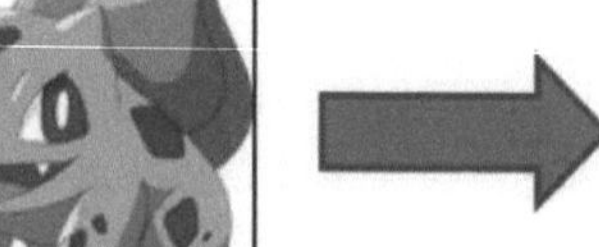
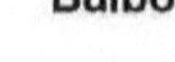

Figure 3: Bulbasaur Available at: http://universoanimanga.blogspot.com.br/

Figure 4: Bulb Available at: http://flores.culturamix.com

Species: Seed Pokémon

Type: Plant/poisonous

Description: One of the first three Pokémon, on its back it carries a bulb, still closed, which nourishes it with solar energy (through photosynthesis). This Pokémon has the characteristics of bulbous plants. Bulbasaur evolved into: 002 - Ivysaur and 003 - Venusaur.

| **043 – Oddish** | **Cebola** |

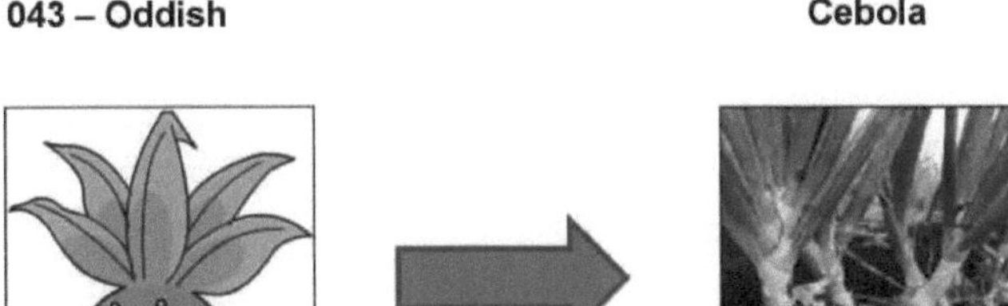

Figura 5:   Oddish Available at: http://universoanimanga.blogspot.com.br/

Figura 6:   Onion Available at: http://www.vegetall.com.br

Species: Leaf Pokémon

Type: Plant/poisonous

Description: It looks like an onion, with leaves on its head. The Oddish evolved into: 044 - Gloom and 045 - Vileplume.

| **069 – Bellesprout** | **Planta Carnívora** |

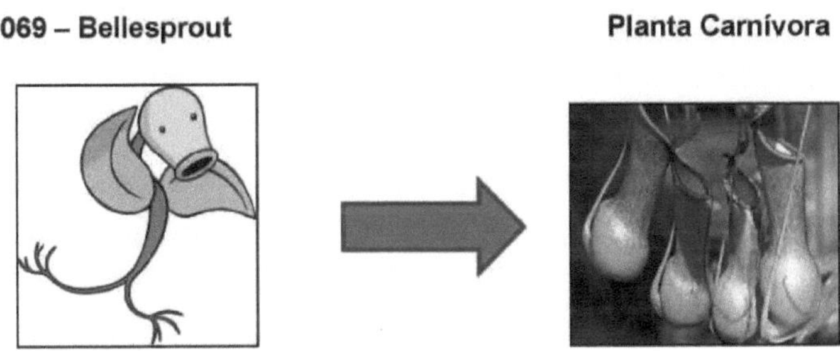

Figure 7: Bellesprout Available at: http://universoanimanga.blogspot.com.br/

Figure 8: Carnivorous plant Available at: http:// cultiveplantascarnivoras.blogspot.com

Species: Flower Pokémon

Type: Plant/poisonous

Description: Uses its roots and vines to fight. The Bellsprout has evolved into: 070 - Weepinbell and 071 - Victreebel.

103 – Exeggutor

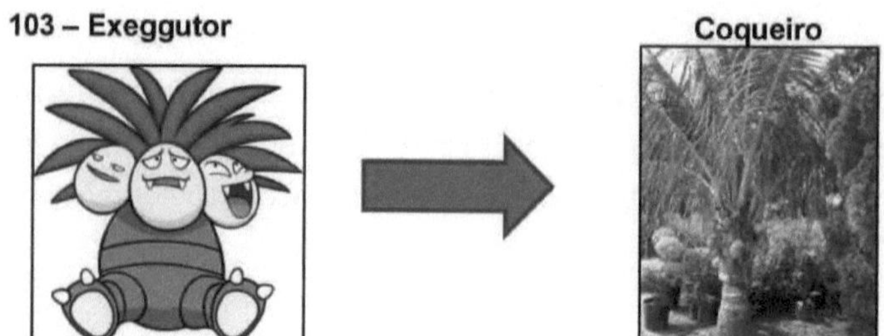

Figure 9: Exeggutor Available at: http://universoanimanga.blogspot.com.br/

Figure 10: Coconut tree Available at: http:// humbertolus.mercadoshops.com.br

Species: Coconut Pokémon

Type: Plant/psychic

Description: It looks like a coconut tree. The Exeggutor is the evolution of the 102 - Exeggcute (several eggs).

114 – Tangela

Figura 11:   Tangela Available at: http://universoanimanga.blogspot.com.br/

Figura 12:   Vine Available at: http:// encostadavila.pt

Species: Vine Pokémon

Type: Plant

Description: Their main tactic is to attack and trap their enemies with their vines of hair, which cover almost their entire body.

Note: Tangela has no evolution.

- Second Generation

From Pokémon 152 to 251 ; The plant characters are:

152 – Chikorita   Folha

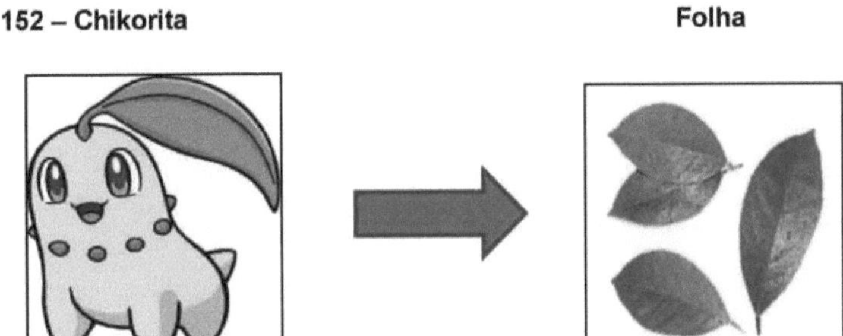

Figure 13: Chikorita Available at: http://universoanimanga.blogspot.com.br/

Figure 14: Vineyard Available at: http:// www.dreamstime.com

Species: Leaf Pokémon

Type: Plant

Description: Very docile, but powerful. The Chikorita evolved to: 153 -

Bayleef and 154 - Meganium.

182 – Bellossom   Flor

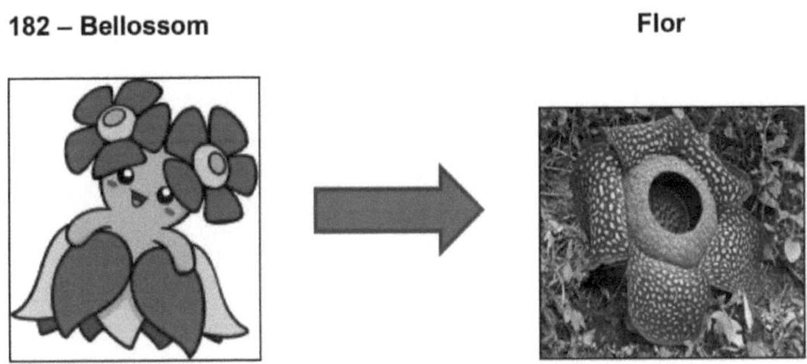

Figure 15: Bellossom Available at: http://universoanimanga.blogspot.com.br/ Figure 16: Vineyard Available at:
http://topbiologia.com

Species: Flower Pokémon **Type:** Plant **Description:** Gloom's second form of evolution,
parallel to Vileplume, it looks like the world's largest single flower. It has such beauty, it emits
an unpleasant smell. Native to the tropical forests of Indonesia, it has an unbearable smell,
similar to a decomposing body, which scares anyone away. The flower is parasitic and has
no roots, leaves or stems.

187 – Hoppip   **Erva Daninha**

Figure 17: Hoppip Available at: http://universoanimanga.blogspot.com.br/ Figure 18: Weed Available at: http://flores.culturamix.com

Species: Cotton Weed Pokémon

Type: Plant/Flyer

Description: It uses its lightness and the wings on its back to fly with the wind. The Hoppip evolved into: 188 - Skiploom and 189 - Jumpluff.

191 – Sunkern   **Semente**

Figure 19: Sunkern Available at: http://universoanimanga.blogspot.com.br/

Figure 20: Seed Available at: http:// www.heladinutri.com

Species: Seed Pokémon

Type: Plant

Description: It's useless in battle. It only collects dew and nutrients. The Sunkern evolved into: 192 - Sunflora.

251 – Celebi   Disseminada pelo vento

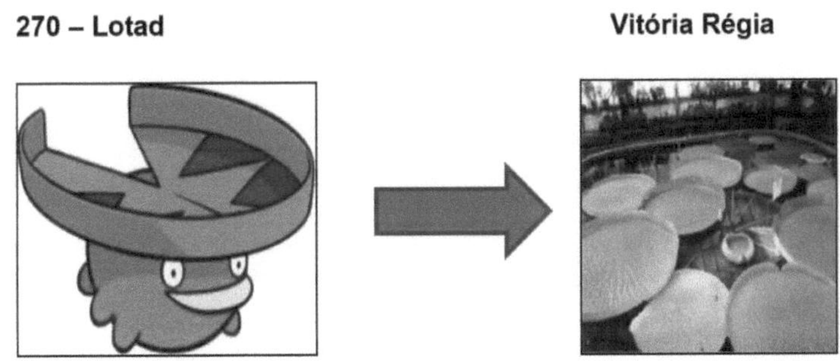

Figure 21: Celebi Available at: http://universoanimanga.blogspot.com.br/ Figure 22: Spread by the wind Available at: <u>http://br.pinterest.com</u>

Species: Time-traveling Pokémon **Type:** Plant/Psychic **Description:** This is a Legendary Pokémon. It lives in forests and has the power to travel through time.

∀ **Third Generation**

From Pokémon 252 to 386; The plant Pokémon are:

270 – Lotad   Vitória Régia

Figure 23: Lotad Available at: http://universoanimanga.blogspot.com.br/ Figure 24: Vitória Régia Available at: <u>http://brasilescola.uol.com.br</u>

Species: Watergrass Pokémon

Type: Plant/Water

Description: Lives submerged in water, leaving only its leaf sticking out. The Lotad evolved into: 271 - Lombre and 272 - Ludicolo.

273 – Seedot   **Noz**

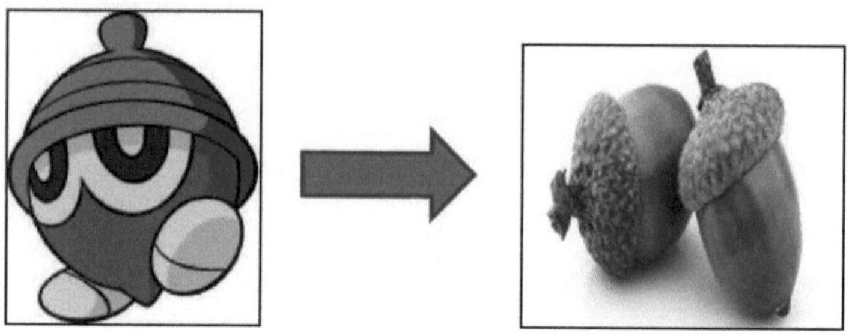

Figure 25: Seedot Available at: http://universoanimanga.blogspot.com.br/

Figure 26: Walnut Available at: http://www.motherjones.com

Species: Acorn Pokémon

Type: Plant

Description: Looks like a walnut. It sticks the tip of its head into tree branches to absorb nutrients. The Seedot evolved into: 274 - Nuzleaf and 275 - Shiftry.

285 – Shroomish   **Cogumelo**

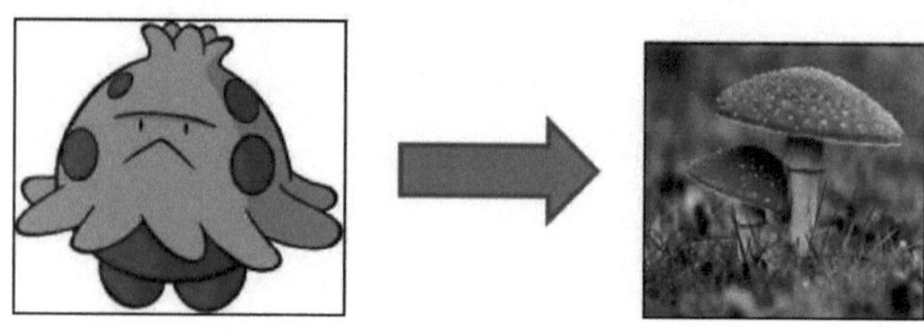

Figura 27:   Shroomish Available at: http://universoanimanga.blogspot.com.br/

Figura 28:   Mushroom Available at: http://www.e-farsas.com

Species: Mushroom Pokémon

Type: Plant

Description: When it feels threatened, it releases poisonous spores from its body. Shroomish evolved into: 286 - Breloom.

357 – Tropius

Fruto

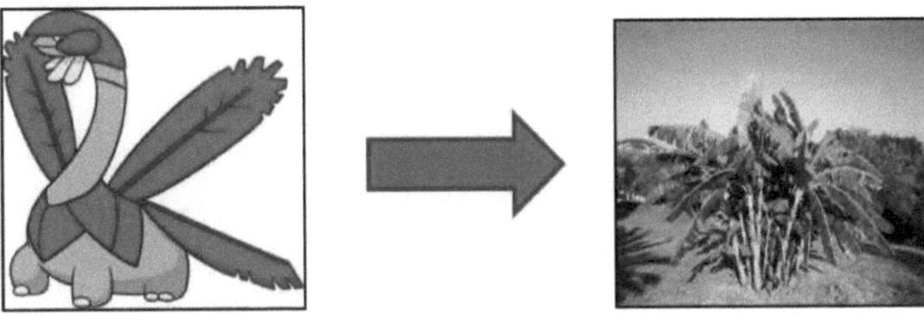

Figure 29: Tropius Available at: http://universoanimanga.blogspot.com.br/

Figure 30: Fruit Available at: http://www.pensamentoverde.com.br

Species: Fruit Pokémon

Type: Plant/Flyer

Description: It is a walking tree that flies by flapping its leaves and can even produce fruit from its own body. Note: It has no evolution.

- Fourth Generation

From number 387 to 493; The plant Pokémon are:

406 – Budew

Broto

Figura 31:   Budew Available at: http://universoanimanga.blogspot.com.br/

Figura 32:   Sprout Available at: http://agroecologia.org.br

Species: Pokémon Sprout

Type: Plant/Venom

Description: It acts like an ordinary plant bud, even releasing pollen. Budew evolved into: 407 - Roserade.

420 – Cherubi   **Cereja**

Figura 33:   Cherubi Available at: http://universoanimanga.blogspot.com.br/

Figura 34:   Cherry Available at: http://dicassobresaude.com

Species: Cherry Pokémon

Type: Plant

Description: The little ball on its head contains nutrients that are absorbed little by little to help it evolve. Cherubi evolved into: 421 - Cherrim.

455 – Carnivine   **Carnívora**

Figure 35: Carnivine Available at: http://universoanimanga.blogspot.com.br/

Figure 36: Carnivora Available at: http://diariosintonia.com.br

Species: Pokémon carnivorous plant

Type: Plant

Description: Hangs from tree branches and keeps its mouth open to attack passing prey. It has no evolution.

492 – Shaymin   **Planta com flor**

Figure 37: Shaymin Available at: http://universoanimanga.blogspot.com.br/

Species: Pokémon Gratidao

Type: Plant

Description: Legendary Pokémon. It can absorb poisons with its flowers and inspires gratitude wherever it goes.

- Fifth Generation

From number 494 to 649; The plant Pokémon are:

546 – Cottonee   **Algodão**

Figura 39:   Cottonee Available at: http://universoanimanga.blogspot.com.br/

Figura 40:   Cotton Available at: http://www.agricultura.gov.br

Species: Cotton Breath Pokémon

Type: Plant

Description: It releases various fragments of cotton from its body to distract its opponents. Cottonee evolved into: 547 - Whimsicott.

548 – Petilil

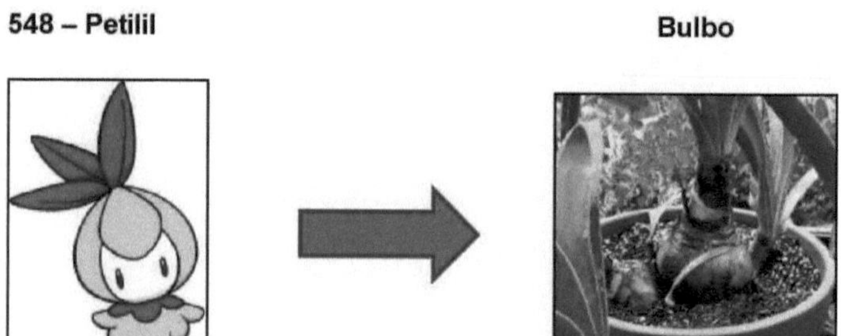

Bulbo

Figura 41:   Petilil Available at: http://universoanimanga.blogspot.com.br/

Figura 42:   Bulb Available at: http:// elicriso.it/

Species: Bulb Pokémon

Type: Plant

Description: It has leaves on its head that can restore the energy of those who eat them, although they are bitter. Petilil evolved into: 549 - Lilligant.

556 – Seismitoad

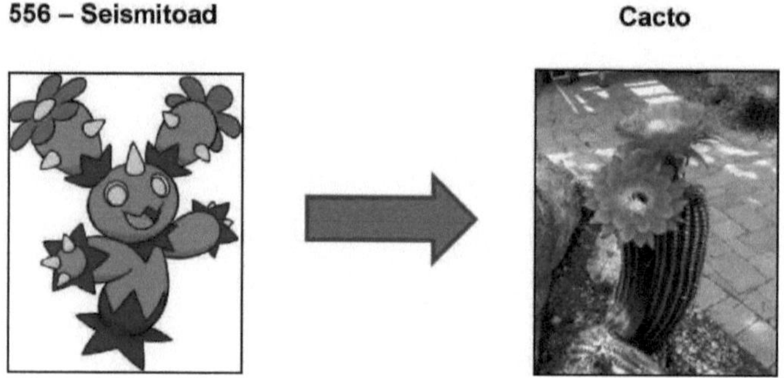

Cacto

Figura 43:   Seismitoad Available at: http://universoanimanga.blogspot.com.br/

Figura 44:   Cactus Available at: http://pinterest.br

Species: Pokémon cactus

Type: Plant

Description: They live in deserts and shake their limbs like maracas.

- Sixth Generation

It starts at number 650 and continues to launch and where it was analyzed up to the 737.

The plant Pokémon are:

650 – Chespin   **Folhas**

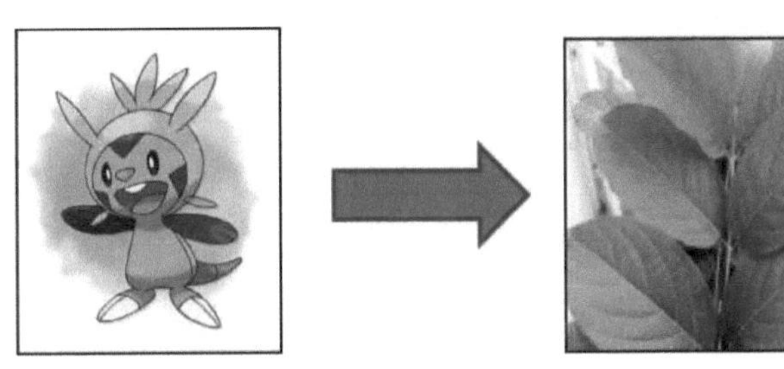

Figure 45: Chespin Available at: http:// pokemonthecks.blogspot.com.br

Figure 46: Sheets Available at: http:// 40forever.com.br/

Species: Leaf Pokémon

Type: Plant

Description: He has a hard shell that protects his head and back. Despite having a curious nature that tends to get him into trouble, Chespin keeps an optimistic outlook and doesn't worry about small details. Chespin evolved into: 651 - Quilladin and 652 - Chesnaught.

Some Pokémon, although described as plants, had more animal than plant characteristics, and it was not possible to identify similar plant species. It is possible that these "little monsters" are hybrids of animals and plants and are the result of the author's imagination. For this reason, these Pokémon were not included in this work.

4.2 Student results

A questionnaire with 15 closed questions was administered to 35 students in the 7th year of elementary school at Colégio Souza Marques. The data obtained in this study, using a questionnaire with closed questions, is shown in the graphs below.

The age of the students interviewed ranged from 11 to 14, as shown in Graph 1. The most prevalent age was 12 (23 students, 65.7%) which, according to school standards, corresponds to the ideal range for the grade in question.

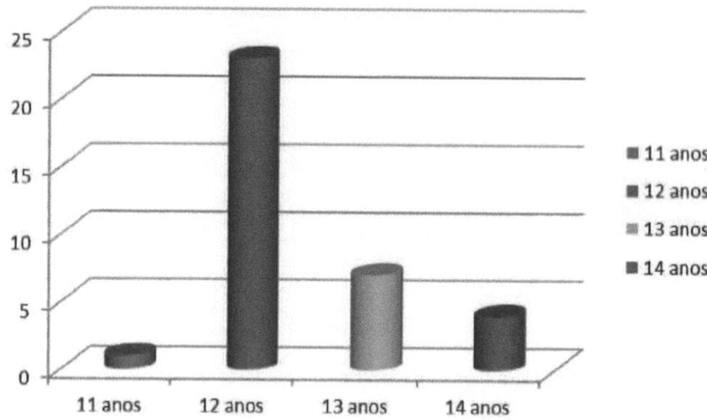

Graph 1. Age group of students at Colégio Souza Marques in the 7th year of elementary school II.

Regarding the presence of a TV at home, 100% of the students have at least one device at home, so they have access to the Pokémon cartoon (Graph 2).

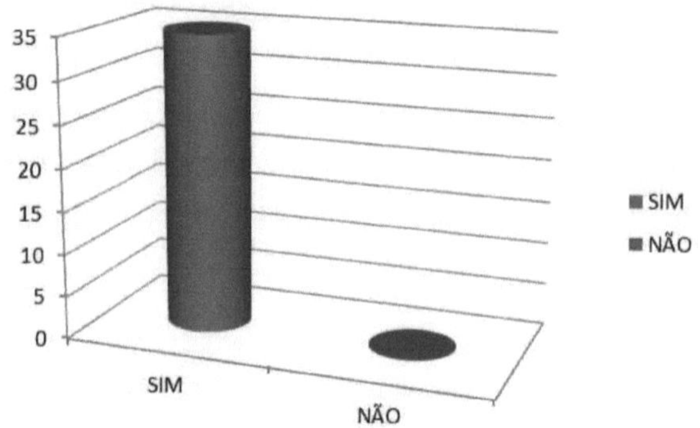

Graph 2. Students who own a TV at home.

This cartoon was shown on both open and closed TV. When asked about this media, 32 students said they had pay-TV and only 3 did not. Therefore, they all had access to the cartoon covered in this study (Graph 3).

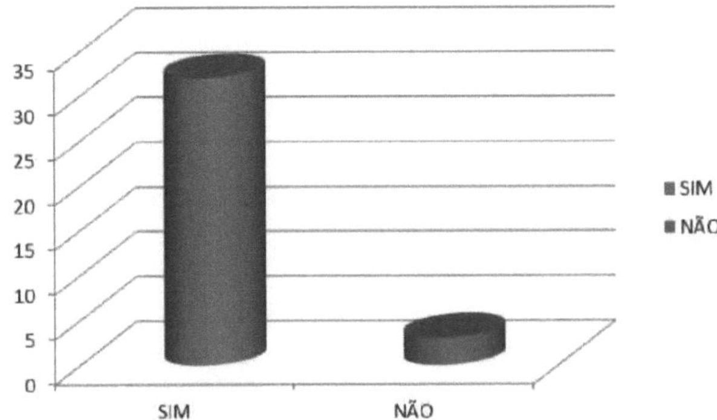

Graph 3. Students who own pay-TV.

When asked about this media, 34 students said they had the internet and only 1 did not. Therefore, everyone can access the cartoon covered in this study, either out of curiosity or because they like the animation (Graph 4).

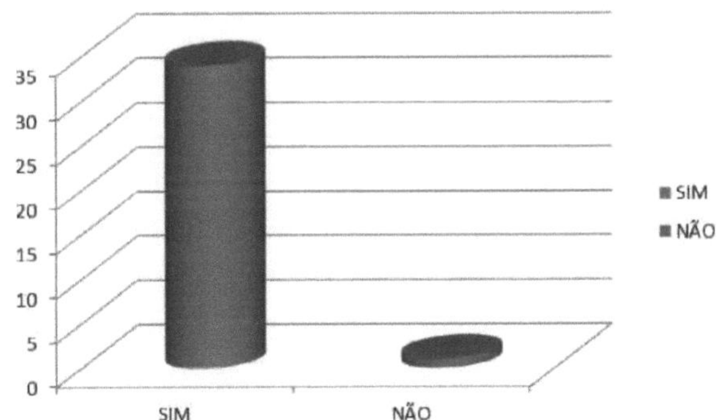

Graph 4. Students who have internet at home.

When asked how often they watch cartoons, 10 students replied that they watch them very often, 23 infrequently and 2 not at all. As such, it can be seen that the vast majority of children today are not so curious or even willing to watch cartoons. According to this data, the growing abundance of technology available to them today and entertainment that can attract them more easily (Graph 5).

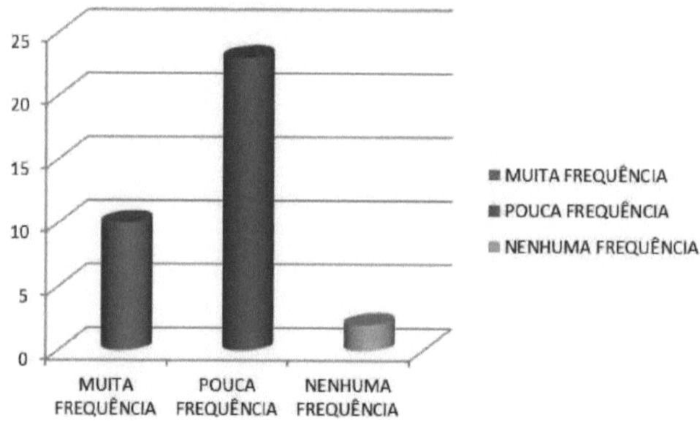

Graph 5. Frequency with which students watch cartoons.

When asked about the influence of drawings on their daily lives, 10 students said there was a lot of influence, 18 students very little and 7 students not very much. Thus, we can compare again that drawings have lost more and more space in children's lives, with technology having a total influence in the 21st century (Graph 6).

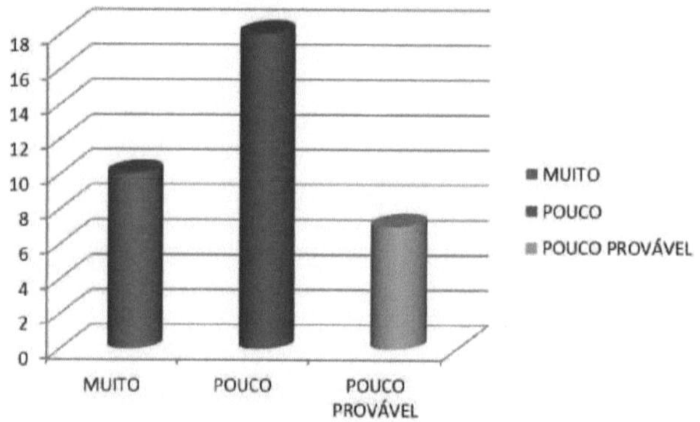

Graph 6. Influence of cartoons on students' daily lives.

When asked if they had ever watched or do watch Pokémon, 25 students answered "yes" and 10 students "no". The cartoon was very popular a few years ago and still is, and most of them have had the opportunity to watch it (Graph 7).

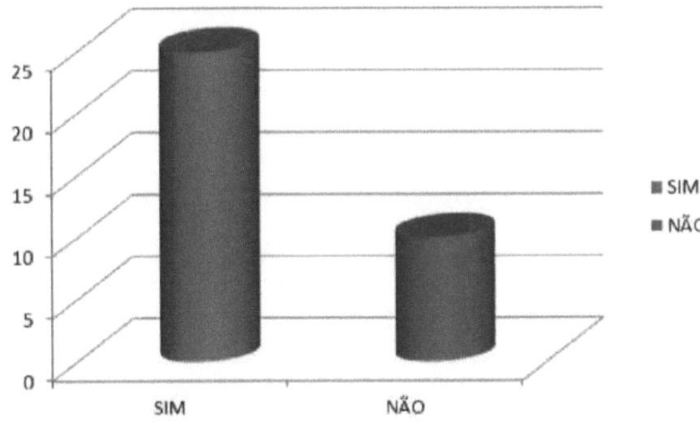

Graph 7. Students who have seen the Pokémon cartoon?

When asked about science teaching and whether it would be influenced by animations containing plants to approach botany in the classroom, 5 students said they felt intrigued, 18 said they felt more stimulated, 6 students said there should be no influence, 5 students didn't care and 1 student didn't answer. Therefore, the use of drawings such as Pokémon, among others, could be used to stimulate students more in class and outside of class. The graph shows a strategy to get the class more interested in the content presented (Graph 8).

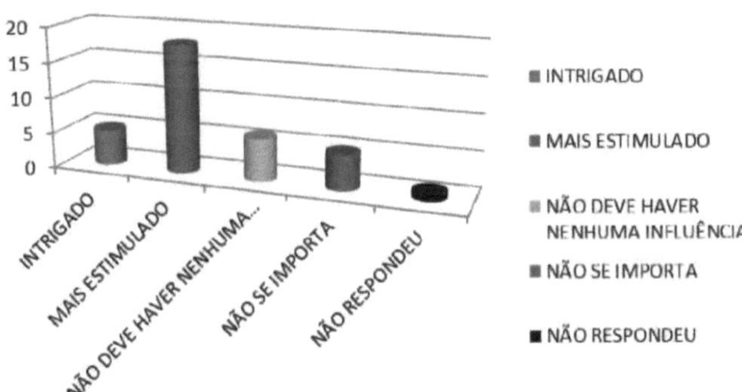

Graph 8. Influence of animations containing plants on interest in studying botany.

When asked if the Pokémon cartoon used plant characters, 29 students answered "yes" and 6 students "no". In botany, drawings can become a means of teaching children about the plant kingdom, becoming a tool for better absorption of the content applied in class. In addition to textbooks, drawing could increasingly stimulate children in the classroom (Graph 9).

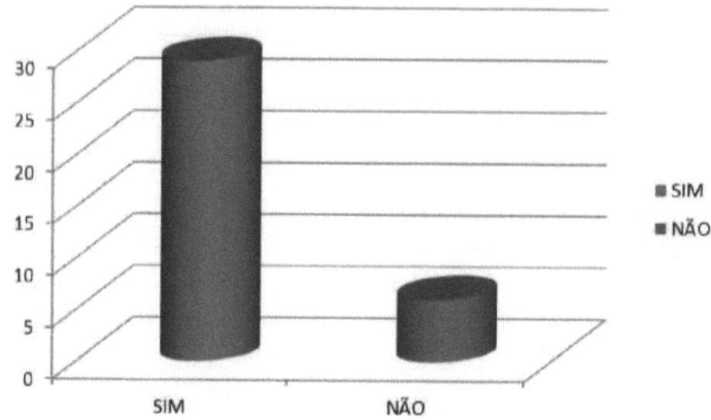

Graph 9. Students who identify the use of plants in Pokémon characters.

When asked if the Pokémon Sunflora is a real plant, 30 students answered correctly and 5 were wrong. Sunflora looks like a sunflower, and the majority were able to understand and distinguish this similarity without any difficulty (Graph 10).

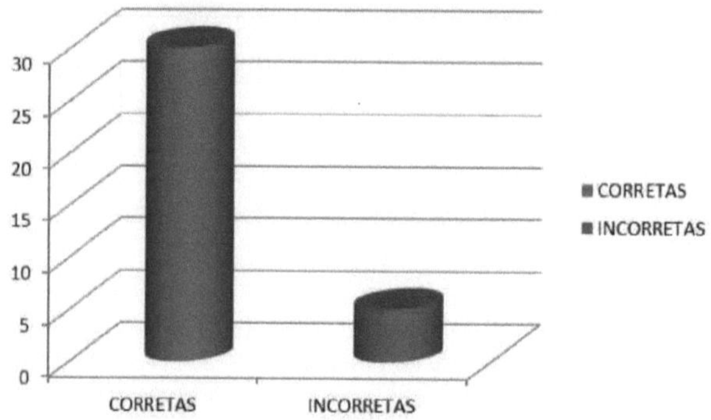

Graph 10 - Students who identified a plant in Pokémon Sunflora.

When asked about the process of photosynthesis, which is the absorption of sunlight and carbon dioxide and the release of oxygen into the environment, using the image of the Pokémon Sunflora because it uses photosynthesis to stay strong during battles, 32 students answered correctly and 3 got it wrong. The question showed that they had learned the content and were able to discuss it in the questionnaire (Graph 11).

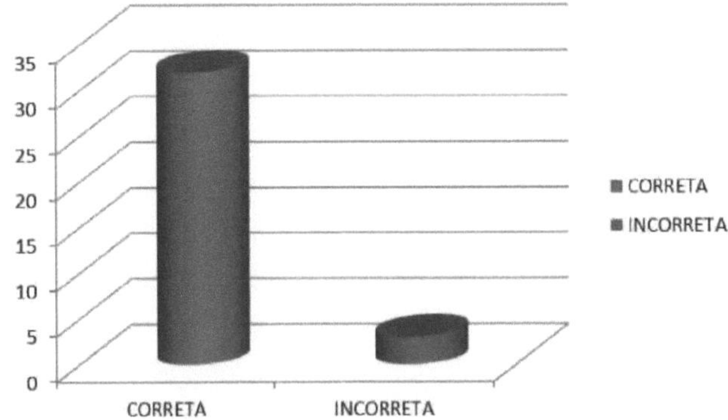

Graph 11 - Students who identified the process of photosynthesis in the physiology of the Sunflora character.

The question about the character Exeggutor being a plant like a coconut tree and its three heads looking like coconuts (fruit), where 24 students answered correctly and 11 students got it wrong. Even though it was easy to interpret and identify, some students had doubts, marking alternatives that were far from what they expected (Graph 12).

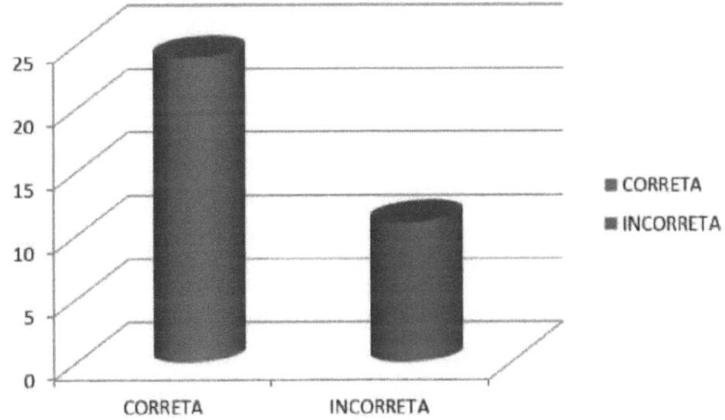

Graph 12: Students who identified the plant represented by the Pokémon Exeggutor

When asked if any drawing had ever been used as a teaching tool and that drawing could be a facilitator of learning, 18 students answered yes, 6 students yes and no, 1 student no and 10 students no and yes. The use of drawings and animations was used in class and facilitated understanding of the subject and could influence learning (Graph 13).

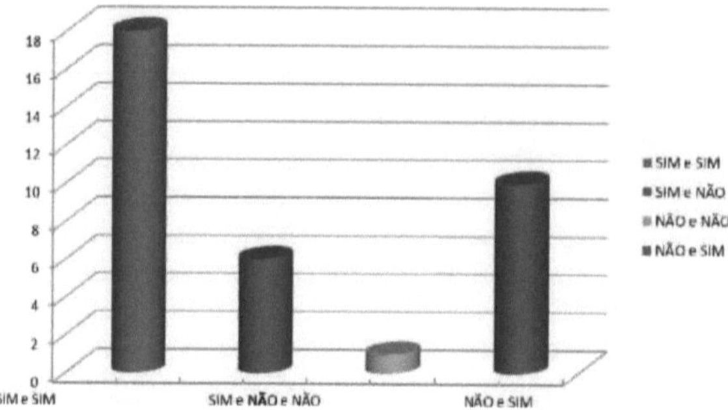

Graph 13. Did the students think that drawing had been used as a teaching tool? Did they think that drawing could facilitate their learning?

Asked if the character Vileplume could be a plant because it has a huge flower on its head and is poisonous, 26 students answered correctly and 9 were wrong. Vileplume bears a resemblance to a plant of the genus Rafflesia and is known for producing the largest flower in the world. The plant is also known for the strong smell it gives off, equivalent to that of decaying meat, which has earned it the nickname of Cape Flower or Monster Flower (Graph 14).

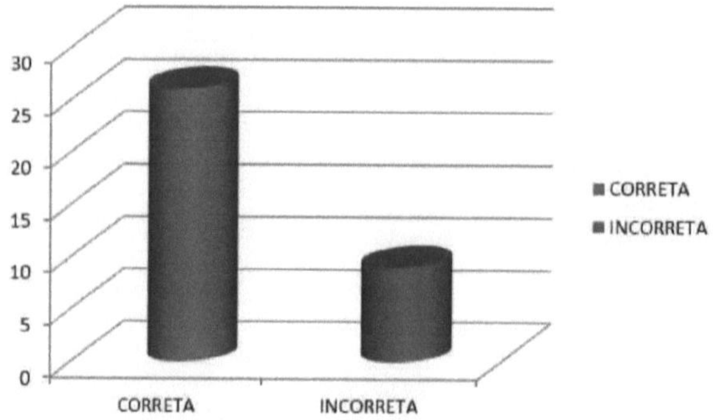

Graph 14. The students talked about Vileplume, a poisonous Pokémon with something that looks like a huge flower on its head. What if they could identify it as a plant?

When asked about the character Victreebel, who has a liquid inside him that attracts enemies, and whether there is a plant with this power, 29 students answered yes and 6 students no. From the Nepenthaceae family, a carnivorous plant that is very similar to Victreebel, has a liquid to attract its prey (Graph 15).

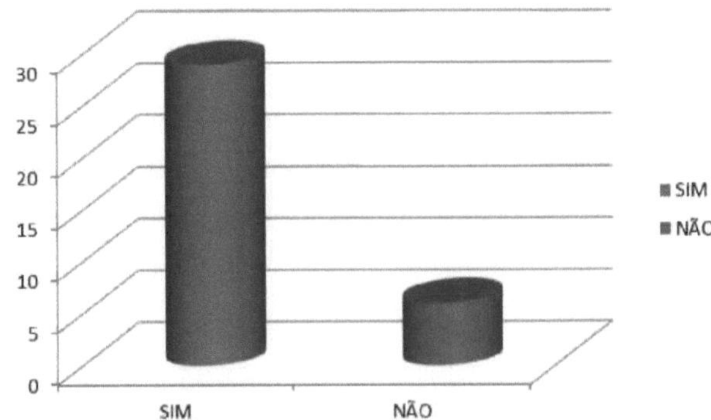

Graph 15. Questionnaire asking elementary school students at Colégio Souza Marques if they know that Victreebel is a Pokémon and has a liquid inside that attracts enemies, do they think there is a plant with this power?

When asked whether the character Jumpluff was a flying Pokémon, 32 students answered correctly and 3 students got it wrong, thus demonstrating their mastery of the subject matter. Jumpluff is similar to anemochoric plants (fruits) that disperse their seeds away from their parent plants (Graph 16).

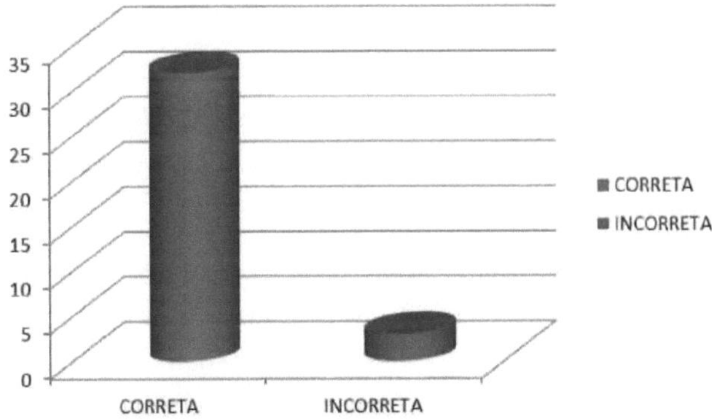

Graph 16. Students who recognized Jumpluff as a flying Pokémon.

5. FINAL CONSIDERATIONS

The purpose of this work was to verify the influence of cartoons in elementary school II, to observe the characteristics that each Pokémon plant has, analyzing the characteristics of each one and identifying its link with reality. By means of a questionnaire applied in the classroom, we observed that 80% of the students were able to identify the similarity between Pokémon and plants.

Therefore, the presence of the game as a facilitator of learning shows that there can be more interaction between the class and thus fulfilling the purpose of having better acceptance of the botany content in class. Indicating this type of tool as an activity so that science has more and more subterfuges to be better accepted by students.

Through the questionnaire, it was possible to see that the concepts that the students had built up about botany, together with the cartoon in question, had brought some benefits to the students' learning, making them feel more at ease when discussing a subject that in the literature is dull and difficult. There was a contribution from both the didactic content and the drawing, as the students were mostly able to identify the characteristics of real plants.

The Pokémon cartoon has its positive points in terms of learning, as it is easy to access and understand for children, most of whom are in the 12-year-old age group, causing interest, curiosity and better interaction with the subject, as it is something current and attractive, even more so with the fever of the Pokémon GO game, which has enabled greater interaction, as it is an augmented reality game and has made it possible for many people to want to be a Pokémon master and go around capturing their "little monsters".

However, it does have its downsides, when the child can become confused with the characteristics covered and without the teacher's help to clarify any doubts or difficulties observed during the lessons, as well as the fact that the child doesn't like drawings, as can be seen from some of the answers in the questionnaire or from the Pokémon itself that was used for this work.

From the studies carried out using the drawing in question, even with little visibility in the 7th grade of the Colégio Souza Marques, a considerable percentage of the questions presented on the content were answered correctly. Most of the students don't think that the drawings can influence their lives, but the majority said that they would be encouraged if this and other drawings were used in class in a playful way to help them learn.

Another observation is that some of the plant Pokémon don't look exactly like plants, many have more pronounced animal characteristics with mere plant details, thus creating confusion and only being able to perceive the reality as a plant many times during their evolution.

With the correct use of the media, we can observe a better interaction between the students, because they feel more familiar with the content covered and have something to draw their attention to in order to better absorb what has been applied.

6. BIBLIOGRAPHICAL REFERENCES

The history of the Pokémon universe - Available at:

<http://pokemonadvancer.blogspot.com.br/p/historia-do-universo.html>. Accessed on: April 2, 2016.

Pokémon How It Was Created - Available at: <

http://kassdouglas.blogspot.com.br/2010/03/pokemon-como-foi-criado.html>.

Accessed on: April 3, 2016.

ALMEIDA,PAULO NUNES DE. **Playful education: techniques and games**

pedagogical. Sao Paulo, SP: Loyola, 2003.

ARELARO, LISETE REGINA GOMES. - **Primary Education in Brazil: Advances, Perplexities and Trends - 2005**. Available at:

http://www.cedes.unicamp.br. Accessed on: May 5, 2016.

BARROS, LAiSSA - **The advantages of technology in teaching children** - Available at: < http://www.b9.com.br/38376/tech/as-vantagens-da-tecnologia-   no-ensino-das-criancas/>. Accessed on: June 6, 2016.

BANTI, R.S. - **The use of comic strips in the teaching of science and biology**. - Mackenzie Presbyterian University - Center for Biological and Health Sciences. Monograph. Sao Paulo, 2012.

BRAZIL. Ministry of Education and Sports. Department of Education

Fundamental. **Referencial Curricular Nacional para a Educaçâo infantil- Conhecimento de Mundo.**Brasilia, MEC/SEF 1988.

BRAZIL. MINISTRY OF EDUCATION. **Parâmetros curriculares nacionais**: ciências da natureza, matemàtica, e suas tecnologias. Brasilia: Secretaria de Educaçâo Média e Tecnològica, 1998.

BIFON, KAROLINE RAQUEL. - **The Use of Cartoons as a Pedagogical Tool for Children's Learning - Universidade Estadual de Londrina 2012** - Available at   at:
   <

file:///C:/Users/fabic/Downloads/KAROLINE%20RAQUEL%20BIFON%20(1).pdf   >. Accessed on: September 21, 2016.

CASTILHO, MARIA INÊS., RICCI, TRIESTE FREIRE - **The Use of Animations as a**

Motivating Element for Learning - Federal University of Rio Grande do Sul. Available at:< http://www.lume.ufrgs.br/handle/10183/622.> Accessed on: November 14, 2016.

COSTA, ELAINE CRISTINA PEREIRA, BARROS, MARCELO DINIZ MONTEIRO DE. - **Light, camera, action: the use of films as a strategy for teaching Science and Biology** - Oswaldo Cruz Institute - FIOCRUZ - RJ. Available at:< http://www.arca.fiocruz.br/handle/icict/10623>. Accessed on: August 5, 2016.

CORRÊA, LEIDNIZ SOARES & BENTO, RAQUEL MATOS DE LIMA - **THE IMPORTANCE OF THE LUDIC FOR LEARNING IN CHILDHOOD EDUCATION** - Pan-American College of Ji-Paranà. Available at: < http://unijipa.edu.br/media/files/54/54_218.pdf . Accessed on: November 17.

Pokémon Go creator: instant success in 20 years - Available at:< https://pokemongobr.club/2016/07/criador-do-pokemon-go-historia-de- long-success/>. Accessed on: October 14, 2016.

FERREIRA, PATRICIA CARVALHO ROCHA. - **The Use of Games in Learning Cell Biology in the 7th Year of Elementary School** - PRÓ- REITORIA DE PÔS-GRADUAÇÂO, PESQUISA, EXTENSÂO E CULTURA - PROPGPEC - 2015. Accessed on: June 7, 2016.

FIGUEIREDO, JOSÉ ARIMATÉA, COUTINHO, FRANCISCO ÂNGELO. AMARAL, FERNANDO COSTA. - **Teaching Botany in a Science, Technology and Society Approach - 2012** - Available at: <

http://revistapos.cruzeirodosul.edu.br/index.php/rencima/article/viewFile/420/35   3   >. Accessed on: May 6, 2016.

FOSSATTI, CAROLINA LANNER. - **Cinema de Animaçâo: Uma trajetória marcada por inovações** - National Meeting of Media History, Fortaleza, Ce, 2009. Available at: < http://www.ufrgs.br/alcar/encontros-nacionais-   1/encontros- nacionais/7oencontro20091/CINEMA%20DE%20ANIMACa0%20Uma%20trajet oria%20marked%20por%20inovacoes.pdf>. Accessed on: March 30, 2016.

LIBÂNEO, J.C. - **Didactics and the learning of thinking and learning: the Historical-Cultural Theory of Activity and the contribution of VasiliDavydov.** -

RevistaBrasileira de Educaçâo. Goiânia, 2004.

Complete list of all Pokémon - Available at:< http://universoanimanga.blogspot.com.br/2012/10/lista-completa-com-todos-os- pokemons_25.html>. Accessed on: September 2, 2016.

LOUREDO, PAULA. - **DEFINITION OF BOTANICALS** - Available at: < http://brasilescola.uol.com.br/biologia/definicao-botanica.htm>. Accessed on: June 6, 2016.

LUYTEN, SÔNIA. - **Anime** - Available at: < http://www.suapesquisa.com/o_que_e/anime.htm>. Accessed on: June 6, 2016.

Guia da semana - Available at:< http://www.guiadasemana.com.br/tv-e-famosos/galeria/30-desenhos-animados-dos-anos-80-e-90-que-deixaram-   saudade, 2015.> Accessed on: June 4, 2016.

MALAQUIAS, MAIANE.SANTOS; RIBEIRO, SUELY DE SOUZA. - THE **importance of play in the teaching-learning process in childhood development -**   **2013** -
Available at: < https://psicologado.com/atuacao/psicologia-escolar/a-importancia-do-ludico-no- teaching-learning-process-in-the-development-of-childhood>. Accessed on June 7, 2016.

MALUF, ÂNGELA CRISTINA MUNHOZ. **Play, pleasure and learning**.
Petrópolis, RJ: Vozes, 2003.

MAZZOTTI, TARSO BONILHA, OLIVEIRA, RENATO JOSÉ DE. - **Science (s) of education**. Rio de Janeiro, RJ. ed. DP&A, 2000.

MELO, MARIA APARECIDA S. - **Conceptions of Adolescence in Jean Piaget** - Available at: < https://psicologado.com/psicologia-general/human-development/jean-piaget's-conceptions-of-adolescence>.
Accessed on: June 6, 2016.

MELO, C.M.R. - **Playful activities are fundamental to subsidize the process of building knowledge (continuation)**. - Philosophical Information. V.2 noi, p.128-137. 2005.

MENDONÇA, ANNA VALESKA PROCÓPIO DE M. MENDES, JOANA D'ARC UMBELINO. SOUZA, SUELLEN C.C. - **A Reflection on the Influence of Cartoons and the Possibility of Using Them as a Pedagogical Resource**   - Available at: < http://portal.sipeb.com.br/santana/files/2010/08/A-influ%C3%AAncia-dos- cartoons.pdf>. Accessed on: September 21, 2016.

MIRANDA, J.C. - **Science with Humor XXI**. - 2012. Available at: Accessed on 17/07/2014.

Microphone: Radio History -   Available   at:< http://www.microfone.jor.br/historia.htm, 2008>. Accessed on: November 10, 2016.

Build your story - Pokémon creation theory - Available at: <http://minilua.com/monte-a-sua-materia-a-teoria-de-criacao-dos-pokemons- 101/>. Accessed on: April 2, 2016.

The Pokémon episode that sent 700 children to hospital - Available at: < http://www.neogamer.com.br/2011/11/conspiracy-o-episodio-   de-pokemon-que.html>. Accessed on: September 4, 2016.

PATRCIO, MIGUEL.   **-Retroprojecçâo-** Available at: <

http://retroprojeccao.blogspot.   com.   br/2006/12/cinemajaponsumaretrospectiva.ht   ml>. Accessed on: June 4, 2016.

PEREIRA, MIRELLY CRISTINA. RUARO, LAURETE MARIA. - **Media and Child Development:**   **Influences of the**

Organizing Play -   **2009**   - Available at   :<

http://www.pucpr.br/eventos/educere/educere2009/anais/pdf/2062_1398.pdf>.   Accessed on: September 21, 2016.

PIAGET, J. - **The formation of the symbol in the child**. - Rio de Janeiro: Zahar, 1975.

PIAGET, J. **Where is education going?** 7. ed. Rio de Janeiro: José Olympio, 1980.

PIAGET, J. **Seis Estudos de Psicologia**. 24. ed. Rio de Janeiro: Forense Universitària, 2002.

POKÉMON BEST WORLD - **History and Creation** - Available at: < http://pokemon-bestworld.webnode.com.pt/universiversopokemon/historia%20e%20cria%C3%A7%C3%A3o/>. Accessed on June 6, 2016.

Pokémon Thecks -   Available at: <

http://pokemonthecks.blogspot.com.br/2013/03/6-geracao.html>. Accessed on: October 13, 2016.

RIBEIRO, ANA CAROLINE. BATISTA, ALINE DE JESUS. - **A Influência da Midia na Criança / Pré-Adolescente e a Educomunicaçâo Como Mediadora desse Contato -** **2010**   -   Available   at: <http://www.ufrgs.br/alcar/noticias-   dos-nucleos/artigos/A%20INFLUENCIA%20DA%20MIDIA%20NA%20CRIANCA%2 0PREADOLESCENTE%20E%20A%20EDUCOMUNICACaO%20COMO%20M EDIADORA%20DESSE%20CONTATO.pdf. > Accessed on: September 29, 2016.

SILVA, JOAO RODRIGO SANTOS. - **The teaching of botany in the view of Biological Sciences   students**   -   University   of   Sao   Paulo.   Available   at:

<http://www.nutes.ufrj.br/abrapec/viiienpec/resumos/R1021-1 .pdf> Accessed on: March 29, 2016.

SCHECHTMANN, E.;MERTINS, J. M.; VELLOSO, H. M. - **Science Company - 7th grade**. - Saraiva. 2 ed. Sao Paulo, 2012.

all **aboutPokémon-** Available at:

<http://blogcelebi.webnode.com.br/tipos-de-pokemons/planta/>. Accessed on: March 30, 2016.

VYGOTSKY, L. - **The social formation of the mind**. - Sao Paulo: Martins Fontes. 1989.

APPENDICES

APPENDIX 1: Questionnaire

Fundação Técnico – Educacional Souza Marques – FTESM

FACULDADE DE FILOSOFIA CIÊNCIAS E LETRAS – FFCL

Curso de Ciências Biológicas

Questionário para uso no Trabalho de Conclusão de Curso de Fabiana Gama Chimes.

A INFLUÊNCIA DO DESENHO ANIMADO *POKÉMON* NO DESENVOLVIMENTO DA APRENDIZAGEM EM BOTÂNICA NO ENSINO FUNDAMENTAL II

Turma: ______   Data: ___/___/2016.

Idade: ______

1. Você possui Tv em casa?

(   ) SIM

(   ) NÃO

2. Você possui Tv por assinatura?

(   ) SIM

(   ) NÃO

3. Possui internet em casa?

(   ) SIM

(   ) NÃO

4. Com que frequência assiste desenhos animados?

(   ) Muita frequência

(   ) Pouca frequência

(   ) Nenhuma frequência

5. Você acha que os desenhos animados podem influenciar no seu cotidiano?

(   ) Muito

(   ) Pouco

(   ) Pouco provável

6. Você assiste ou já assistiu alguma vez o desenho Pokémon?

( ) SIM

( ) NÃO

7. O ensino de ciências é baseado na transmissão de conhecimentos, como você se comporta com a influência de animações que contenham plantas para abordar a botânica em sala de aula?

( ) Intrigado

( ) Mais estimulado

( ) Não deve haver nenhuma influência

( ) Não se importa

8. Pokémon é uma animação que utiliza personagens planta, você sabia?

( ) SIM

( ) NÃO

9. O Pokémon Sunflora parece uma planta real? Identifique nas altenativas abaixo:

( )

( )

( )

( )

10. O personagem Sunflora possue uma espécie de poder, quando absorve energia solar é capaz de converte-la em energia para manter-se forte durante uma batalha. Isso significa que ele faz um processo biológico importante nas plantas. Que processo é esse?

a) Glicólise

b) Fotólise

c) Fotossíntese

d) Osmorregulação

11. O Pokémon Exeggutor é uma planta, ele tem "três cabeças" que se parecem cocos? Nas figuras abaixo como você identificá-lo?

Exeggutor

( ) 

51

12. Algum desenho já foi utilizado como ferramenta pedagógica? Você acha que o desenho pode ser um facilitador para seu aprendizado?

( ) SIM e SIM

( ) SIM e NÃO

( ) NÃO e NÃO

( ) NÃO e SIM

13. O Pokémon Vileplume é um Pokémon venenoso, com algo que parece uma flor enorme na cabeça. Consegue identifica-lo como uma planta?

Vileplume

( ) 

( ) 

14. Sabendo que Victreebel é um Pokémon e dentro de seu interior possui um liquido em seu interior que atrai inimigos, você acha que existe alguma planta com esse poder?

Victreebel

( ) SIM

( ) NÃO

15. Jumpluff é um Pokémon voador, ele gosta de viajar flutuando com os ventos. Indentifique-o:

Jumpluff

( ) 

( ) 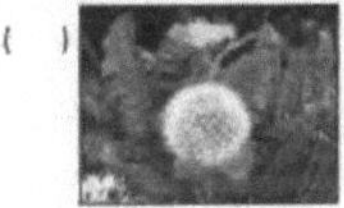

APPENDIX 2: Authorization

FUNDAÇÃO TÉCNICO-EDUCACIONAL SOUZA MARQUES
FACULDADES DE FILOSOFIA, CIÊNCIAS E LETRAS

Ofício no 29/2016

Rio de Janeiro, 15 de setembro 2016

De: Profa. Luciana Cresta de Barros Dolinsky – Coordenadora
Para: Profa Leopoldina de Souza Marques – Diretora

Assunto: **Permissão para Coleta de Dados**

Prezada Profa. Leopoldina,

Solicito autorização para que a aluna Fabiana Gama Chimes possa aplicar, no segundo semestre do ano de 2016, um questionário (em anexo) para os alunos do Ensino Fundamental II do Colégio Souza Marques, a fim de analisar o conhecimento deste grupo sobre a influência do desenho animado no aprendizado de botânica.

Os dados obtidos serão utilizados na confecção do Trabalho de Conclusão de Curso (TCC) de Licenciatura do aluno, orientado pelo professor André Costa Ferreira. O TCC intitula-se: "A Influência do desenho animado Pokemón no ensino de botânica no Fundamental II" e os resultados obtidos poderão ser fornecidos à Instituição, caso solicitado.

Aguardando pronunciamento de V. Sa.,

Atenciosamente,

Professora Luciana C. B. Dolinsky

Coordenadora do Curso de Ciências Biológicas
Fundação Técnico-Educacional Souza Marques

Av. Ernani Cardoso, 335/345 – Cascadura –Rio de Janeiro, RJ – CEP: 21310-310 – Tel.: 2128-4900 Fax: 3350-5983
Site: www.souzamarques.br E-mail: souzamarques.edu/souza@souzamarques.br

Printed by Books on Demand GmbH, Norderstedt / Germany